FA–225

LANDFILL FIRES
THEIR MAGNITUDE, CHARACTERISTICS, AND MITIGATION

MAY 2002

Prepared by

TriData Corporation
1000 Wilson Boulevard
Arlington, Virginia 22209

for

Federal Emergency Management Agency
United States Fire Administration
National Fire Data Center

U.S. FIRE ADMINISTRATION
MISSION STATEMENT

As an entity of the Federal Emergency Management Agency, the mission of the U.S. Fire Administration is to reduce life and economic losses due to fire and related emergencies through leadership, advocacy, coordination, and support. We serve the Nation independently, in coordination with other Federal agencies and in partnership with fire protection and emergency service communities. With a commitment to excellence, we provide public education, training, technology, and data initiatives.

CONTENTS

ACKNOWLEDGMENTS

The United States Fire Administration greatly appreciates the help of the following persons who provided information or reviewed this report:

Rodney Slaughter

Rodney Slaughter is the President of Dragonfly Communications Network, a fire service training and consulting firm.

Todd Thalhamer

Todd Thalhamer, P.E., is a waste management engineer with the California Integrated Waste Management Board. Mr. Thalhamer specializes in investigating and mitigating landfill and tire fires for the State of California.

Dr. Tony Sperling

Dr. Sperling is the President of Sperling Hansen Associates and the founding partner of Landfillfire.com (http://www.Landfill-fire.com). Since 1997, he has specialized in landfill fire risk reduction training and landfill fire extinguishment on more than 30 landfill fire projects.

EXECUTIVE SUMMARY

Landfills can be controversial in and of themselves. Homeowners and business owners tend not to support the siting and development of landfills in their neighborhoods due to perceived notions about noxious fumes, health and environmental effects, and adverse influences on property values. Fires occurring in landfill sites are an ongoing, complex problem that has existed for decades.

Although relatively uncommon, fires in landfills generally receive substantial media attention and have the potential to become politically damaging events. Landfill fires threaten the environment through toxic pollutants emitted into the air, water, and soil.

Landfill fires are particularly challenging to the fire service. A large landfill fire normally requires numerous personnel and a significant period of time before it is contained. Both of these circumstances can strain a jurisdiction, particularly one dependent on volunteer staffing.

Landfill operators, members of the fire service, and community residents need to learn as much as possible from past experience to prevent and mitigate future landfill fires.

REGULATION. In 1976, Congress passed the Resource Conservation and Recovery Act (RCRA), which gave the Environmental Protection Agency (EPA) the authority to control hazardous waste from "cradle-to-grave." RCRA covers the generation, transportation, treatment, storage, and disposal of hazardous waste and provides a framework for the management of non-hazardous wastes. A turning point in landfill regulation and remediation occurred in 1980, first with the "Superfund" legislation, followed by the Hazardous and Solid Waste Amendments (HSWA) in 1984, which finally gave the EPA regulatory authority over landfills. The Comprehensive Environmental Response, Compensation, and Liability Act (CERCLA), known as Superfund, governs closed and abandoned hazardous material waste sites, provides for the liability of persons responsible for the release of hazardous materials at these sites, and established a trust fund to provide for cleanup where no responsible party could be identified.

CHARACTERISTICS. The most common type of landfill is one that is designed to accept municipal solid waste (MSW). Other types of landfills include hazardous materials landfills, construction and demolition landfills, and industrial landfills. Each type of landfill has specific characteristics based on the type of waste it is designed to accept.

The passage of liquid through solid waste in a landfill creates leachate, which contains potentially dangerous pollutants. As such, landfills must operate in a manner that protects the environment, particularly surface and ground waters, from leachate contamination. To do this, landfill designs generally incorporate a composite liner and a leachate collection system, and landfill procedures require that the waste collected each day be completely covered.

Because of the methods normally adopted to deposit, compact, and cover waste in land-fills, the decomposition of waste is largely anaerobic, which results in the production of large quantities of methane and carbon dioxide. Landfills are the largest source of methane emissions in the United States; in 1999, 35 percent of methane emissions were from landfills. Methane is highly flammable and plays a large role in the ignition of landfill fires.

EXTINGUISHING LANDFILL FIRES. The different dynamics, characteristics, and regu-lations of landfills and the fires that occur in them suggest that firefighting tactics need to be de-termined on a case-by-case basis depending on the materials buried in the landfill, which materi-als have ignited, depth of the fire, and the fire's ignition source. Challenges explored in this report include wind/weather; water supply; multi-agency response; personnel safety; access to, access by and maneuverability of heavy equipment; logistics; environmental impact; and landfill contents (potentially hazardous or illegal).

PREVENTION. Fire prevention actions can reduce property damage and the risk of in-jury and death, as well as decrease health and environmental hazards associated with landfill fires. As a rule, the cost of prevention is less expensive than the cost of fighting and cleaning up a fire. In many cases, particularly at larger landfills, fire prevention activities are mandated by law. The principal methods for landfill fire prevention include effective landfill management and ap-propriate methane gas detection and collection.

STATISTICAL ANALYSIS. Data from the National Fire Incident Reporting System (NFIRS) does not include MSW landfills as a fixed property use category. Rather, the NFIRS data set includes a category for "dump or sanitary landfill" under NFIRS Fixed Property Use code 932. Although this definition is broader than the definition of a landfill, it is the closest match available in NFIRS. Based on extrapolation of the NFIRS data, each year in the United States an average of 8,400 dump and landfill fires are reported to the fire service. This represents less than a half percent of all reported fires. Undoubtedly, some landfill fires go unreported because they burned undetected or were on private property and extinguished by the landfill operator. Reported fires are responsible for less than 10 civilian injuries, 30 firefighter injuries, and between $3 and $8 million in property loss each year.[1] Deaths (civilian or fire service) are rare in these fires. Since NFIRS represents a sample of data, it may be that fatalities occurred during the study period and were not reported or captured in the data.

CASE STUDIES. A sample of landfill fires throughout the world sheds light on the land-fill fire problem. Waste disposal practices and the regulation of landfill sites are similar in the comparison countries. Landfill fires have been investigated and studied in more detail in these jurisdictions than in the United States. In addition to presenting U.S. case studies, this report in-cludes brief synopses of interviews and media reports detailing landfill fires in the United States and the lessons that were learned from them.

---

[1] National estimates are based on NFIRS data (1996–1998) and the National Fire Protection Association's (NFPA) annual survey, *Fire Loss in the United States*.

LANDFILL FIRES
THEIR MAGNITUDE, CHARACTERISTICS, AND MITIGATION

Fires occurring at landfill sites across the United States are an ongoing, complex problem that has existed for decades. Landfill fires threaten the environment through toxic pollutants emitted into the air, water, and soil. These fires also pose a risk to firefighters and civilians who are exposed to the hazardous chemical compounds they emit. The degree of risk depends in part on the contents buried in the landfill, the geography of the landfill, and the nature of the fire. There can be great difficulty in the detection and extinguishment of landfill fires, which is compounded because these fires often smolder for weeks under the surface of the landfill before being discovered.

This report was prepared by TriData Corporation, Arlington, Virginia, under contract to the Federal Emergency Management Agency, U.S. Fire Administration (USFA), National Fire Data Center. It presents an overview of the landfill fire problem. Issues examined include the landfill components that create fire hazards; the effect of Environmental Protection Agency (EPA) regulations and landfill cleanup efforts; a profile of landfill fires including their characteristics, methods of extinguishing, and safety issues for firefighters; prevention efforts to reduce landfill fires; and past examples of significant landfill fires and lessons learned.

SOURCES OF DATA

Data on the number of municipal solid waste (MSW) landfill sites in the United States and their current regulations regarding disposal, including those open for disposal and those retired from service, were obtained from the EPA. Data and regulation information pertaining to the Superfund project, including current maps outlining ongoing landfill cleanup efforts, were also obtained from the EPA.

The EPA derives their landfill statistics from *BioCycle* magazine, which conducts an annual survey called "The State of Garbage in America." *BioCycle* magazine sends the survey to state officials and follows up the collected data with phone calls, e-mails, and letters to obtain as complete and accurate information on each participating state as possible. The survey collects data on MSW disposal practices in the United States, including information on national recycling rates, number of municipal solid waste landfills, and disposal rates.

Other information on landfill definitions, landfill dynamics, landfill regulations, and chemical compounds contained in emissions were derived from several sources within the EPA.

Landfill fire statistics presented here are based on data from the National Fire Incident Reporting System (NFIRS). NFIRS, established in 1975, is a data system maintained by USFA and today is the largest fire data set in the world. Not all fire departments participate in NFIRS, but the distribution of participants in NFIRS is reasonably representative of the entire nation, even though the sample is not random. Since the data set is incomplete and represents only a sample of American fire departments (<40 percent), many of the numbers in this analysis are national estimates or percentages rather than raw totals or absolute numbers.

Technical information on the characteristics of landfill fires was gathered from sources ranging from the textbook *The Essentials of Firefighting*[2] to various international studies on landfill fires.

Interviews were conducted with fire department representatives who have dealt with landfill fires. Examples of these fires are included in the report, along with lessons learned by the departments in suppressing the fires. Media reports (newspapers, magazines) provided further information about those fires discussed during the interviews.

WHY STUDY LANDFILL FIRES?

Landfills tend to be controversial in and of themselves. Homeowners and business owners may not be inclined to support new siting or development in their areas due to perceived notions about noxious fumes, health effects, and adverse influences on property values. As such, landfill fires can raise political issues and have implications for elected officials on election day. Further, the costs associated with fire suppression and environmental monitoring during a landfill fire can be enormous. This raises questions as to who is responsible for those costs—the municipal jurisdiction, a private company that operates the landfill, a combination of both, or some other entity.

Although relatively uncommon, fires in landfills generally receive substantial media attention. In some cases, landfill fires can smolder for weeks, producing odorous and noxious smoke that can be a community annoyance and that pose a health risk to civilians, firefighters, and others who are exposed.

Depending on the type of landfill and its contents, the smoke from a landfill fire may contain dangerous chemical compounds, which can cause respiratory disorders and other medical conditions. Even if the smoke is benign, it can still aggravate existing respiratory conditions and reduce visibility around the landfill. In addition, contrary to conventional thinking, the use of large amounts of water to suppress a landfill fire can actually make the fire worse by increasing the rate of aerobic decomposition, which increases the heat available inside the landfill. Further, runoff from suppression efforts can overwhelm a landfill's leachate collection system and contaminate ground or surface water sources.

---

[2] *Essentials of Firefighting 4th Edition*, International Fire Service Training Association, 2001.

Landfill fires are particularly challenging to the fire service. A large landfill fire will generally require numerous personnel and significant amounts of time to contain. Both of these circumstances can strain a jurisdiction, particularly one dependent on volunteer staffing. Depending on the type and location of the fire, extinguishing it may require specialized personnel and equipment that may not be immediately available. For example, fires involving hazardous materials require specially trained personnel who are equipped with specialized protective gear. Underground fires generally necessitate the use of heavy equipment (bulldozers, excavators, etc.) to dig out burning waste to be extinguished. Fire may also compromise the structural integrity of a landfill, posing a collapse hazard for personnel operating on the fireground.

Because these fires are relatively uncommon, it is important for communities and the fire service to learn as much as possible from past experience to prevent and mitigate future landfill fires and, if one occurs, to understand the best methods for extinguishing it.

CHARACTERISTICS OF LANDFILLS

Landfills have a variety of unique characteristics, which are primarily determined by the type of waste they are designed to accept. Landfills are regulated by different agencies at the federal, state, and local levels. (Regulatory mechanisms are discussed in detail later in this report.)

The characteristics of landfills constructed before 1984, however, may not conform to those discussed in this section. Prior to 1984, no federal agency had the jurisdiction to regulate landfills. Although some state-based agencies may have had regulatory authority before then, older landfill sites may have accepted both hazardous and nonhazardous waste if they were in operation prior to federal or state regulation. Further, older facilities may not have been constructed with leachate collection systems, gas-monitoring systems, or composite liners that meet the specifications required today.

MUNICIPAL SOLID WASTE LANDFILL. The most common type of landfill is designed for the disposal of municipal solid waste. MSW includes household waste such as product packaging, food scraps, furniture, clothing, and grass clippings. In 1999 alone, Americans generated nearly 230 million tons of MSW.[3] Table 1 illustrates the components of the MSW produced in 1999 by material category. Only 57 percent of this waste, however, went to a landfill for disposal; the remainder was either recovered through recycling (28 percent) or incinerated (15 percent).[4]

The Code of Federal Regulations (CFR) defines an MSW landfill (MSWLF) as "a discrete area of land or an excavation site that receives household waste, and that is not a land application unit, surface impoundment, injection well, or waste pile...MSWLF unit may also receive other types of RCRA [Resource Conservation and Recovery Act] Subtitle D wastes, such

---

[3] U.S. Code of Federal Regulations, 40 CFR 258.2 (Title 40–Protection of Environment Chapter I–Environmental Protection Agency. Part 258 – Criteria For Municipal Solid Waste Landfills).

[4] *Municipal Solid Waste Basic Facts*, Environmental Protection Agency, Office of Solid Waste, January 4, 2002. http://www.epa.gov/epaoswer/non-hw/muncpl/facts.htm.

Table 1. Components of MSW Produced in 1999
(prior to recycling)[5]

Component	Percent of Waste
Paper	38.1
Yard Waste	12.1
Food Waste	10.9
Plastics	10.5
Metals	7.8
Rubber, Leather and Textiles	6.6
Glass	5.5
Wood	5.3
Other	3.2

as commercial solid waste, nonhazardous sludge, conditionally exempt small quantity of genera-tor waste and industrial solid waste. Such a landfill may be publicly or privately owned."[6]

The passage of liquid through the solid waste in a landfill creates leachate. Leachate is defined as "a liquid that has passed through or emerged from solid waste and contains soluble, suspended, or miscible materials removed from such waste."[7] As such, MSW landfills must oper-ate in a manner that protects the environment, particularly surface and ground waters, from leachate contamination. To do this, MSW landfills generally use a combination of a composite liner and a leachate collection system. A composite liner "combines an upper liner of a synthetic flexible membrane and a lower layer of soil at least 2 feet thick with a hydraulic conductivity of no greater than 1×10^{-7} cm/sec"[8] (Figure 1). A leachate collection system consists of a network of pipes that collect the leachate. The collected leachate is typically pumped to the surface of the landfill so that it can be treated and decontaminated. "The leachate collection system must be designed to keep the depth of the leachate over the liner to no greater than 30 centimeters."[9]

While an MSW landfill is in operation, waste is disposed of in layers. These layers are compacted to the smallest practical volume and covered with earthen material at the end of each operating day, except at facilities exempt from cover placement or that use an alternate daily cover such as a tarp.

When a landfill reaches its capacity for waste disposal, a final cover is constructed. The final cover must be designed and constructed to minimize the flow of water into the closed land-fill. It must also contain an erosion layer to prevent the disintegration of the cover. This layer must be composed of a minimum of 6 inches of earthen material capable of sustaining plant

---

[5] *Municipal Solid Waste in the United States: 1999 Facts and Figures*, Environmental Protection Agency.

[6] U.S. Code of Federal Regulations, 40 CFR 258.2 , op. cit.

[7] Ibid.

[8] *Criteria for Solid Waste Disposal Facilities: A Guide for Owners/Operators*, Environmental Protection Agency, EPA/530-SW-91-089, March 1993.

[9] Ibid.

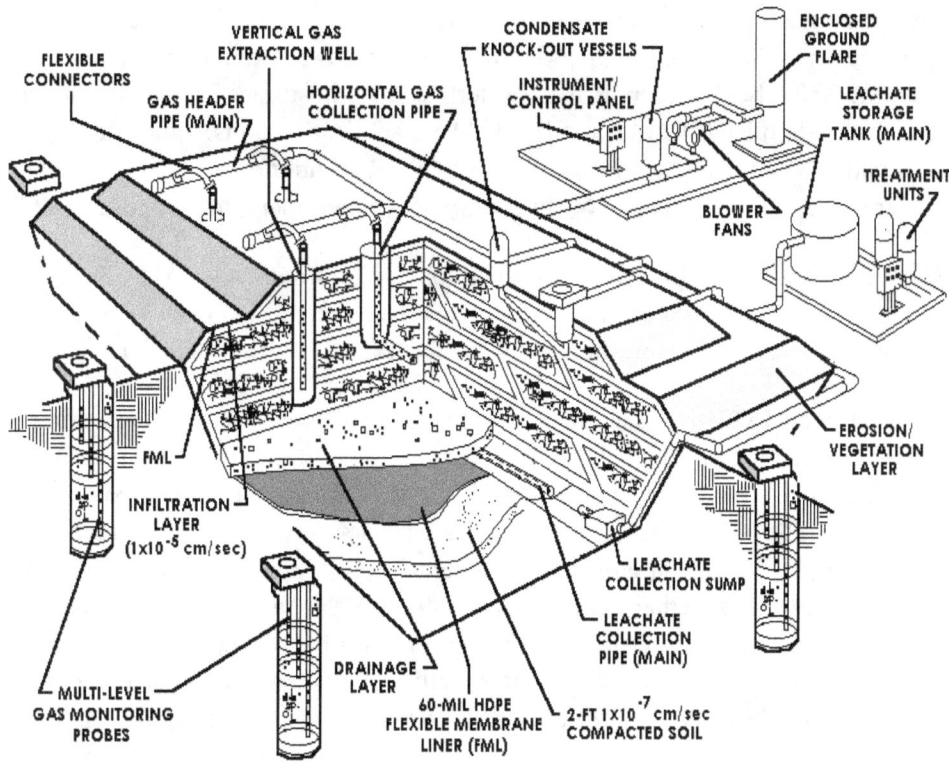

Figure 1. Landfill Components[10]

growth. An independent engineer must certify that the landfill was closed in accordance with federal regulations. For the next 30 years, landfill owners or operators are required to maintain the integrity of the final cover, monitor groundwater and methane gas, and continue leachate management. Finally, the property deed must reflect the property's prior use as a landfill, which restricts the future development of the site.[11]

OTHER TYPES OF LANDFILLS. Some types of waste (e.g., industrial waste and hazardous waste) cannot necessarily be disposed of in an MSW landfill. Instead, these materials must be disposed of in specially designed landfills or in MSW landfills in limited quantities.

Construction and Demolition. Waste from construction and demolition (C&D) projects, including untreated lumber, drywall, plaster, plumbing materials, etc., is not considered MSW. These wastes can be deposited either in MSW landfills or in specially constructed C&D landfills that are required to meet less stringent regulations than MSW landfills. Based on anecdotal remarks by landfill fire suppression professionals, C&D landfills are at a much higher risk for a significant fire than other types of landfills.[12]

---

[10] Courtesy of the California Integrated Waste Management Board.

[11] *Criteria for Solid Waste Disposal Faciliities,* op. cit.

[12] From information received in e-mail correspondence with Dr. Tony Sperling, P.Eng.

Industrial. Each year, about 7.6 billion tons of industrial waste are generated and managed by manufacturing facilities. The majority of this waste is wastewater or non-wastewater sludges and solids. Nearly 97 percent is wastewater managed in surface impoundments; the remainder is managed in landfills, waste piles, and land application units.[13] Industrial waste is classified as neither MSW nor hazardous waste under RCRA Subtitle C, which places industrial landfills under the regulatory authority of states and local government, not the federal authorities.

Hazardous Materials. In 1999, 1.4 million tons of hazardous waste were disposed of in landfills.[14] Hazardous waste landfills are similar in character and design to MSW landfills, but they are required to meet more stringent regulations for leachate collection and decontamination.

LANDFILL EMISSIONS. Landfill emissions are the result of the decomposition of organic materials in the landfill (including yard waste, household waste, food waste, and paper). Because of the nature of the construction of landfills, this decomposition is anaerobic[15] and results in the production of large quantities of methane (which is highly flammable) and carbon dioxide. In fact, landfills are the largest source of methane emissions in the United States, accounting for 35 percent of methane emissions in 1999.[16] MSW landfills generate about 93 percent of U.S. landfill emissions; industrial landfills account for the remaining emissions.[17] Methane emissions from landfills are affected by site-specific factors such as waste composition, available moisture, and landfill size.[18] Approximately 28 percent of the methane generated in landfills in 1999 was recovered.[19] The remainder of landfill-generated methane was dispersed in the air.

Approximately 50 percent of gas emitted from landfills is methane; carbon dioxide accounts for about 45 percent, and the remainder is composed of nitrogen, oxygen, hydrogen, and other gases.[20] Both methane and carbon dioxide are greenhouse gases that pose environmental problems. Of the two gases, methane is far more potent than carbon dioxide. Methane has a global warming potential (GWP)[21] of 21 over a 100-year period. This means that on a kilogram-for-kilogram basis, over a 100-year period, methane is 21 times more potent than carbon dioxide in causing climate change.[22]

---

[13] *Guide for Industrial Waste Management*, Environmental Protection Agency, EPA530-R-99-001, June 1999.

[14] *National Biennial RCRA Hazardous Waste Report*, Environmental Protection Agency, EPA530-S-01-001, June 2001, p. ES-8.

[15] An *anaerobe* is an organism, such as a bacterium, that can live in the absence of atmospheric oxygen. Conversely, an *aerobe* is an organism that requires oxygen to live.

[16] *Inventory of U.S. Greenhouse Gas Emissions and Sinks,* Environmental Protection Agency, EPA 236-R-01-001, April 2001, p. ES-19.

[17] U.S. Methane Emissions 1990-2000: Inventories, Projections, and Opportunities for Reductions, Environmental Protection Agency, EPA 430-R-99-013, September 1999 , p. 2-1.

[18] Inventory of U.S. Greenhouse Gas Emissions and Sinks, op. cit.

[19] Ibid.

[20] *Landfill Methane Outreach Program*, Environmental Protection Agency, FAQ Sheet, June 2001.

[21] The term *global warming potential* has been developed by the EPA to compare the ability of each greenhouse gas to trap heat in the atmosphere relative to another gas. This measurement of GWP relies on carbon dioxide as the reference gas. The GWP of a greenhouse gas is the ratio of global warming (both direct and indirect) from one unit mass of a greenhouse gas to one unit mass of carbon dioxide over a set period of time.

[22] Climate Change, Methane and Other Greenhouse Gases, Environmental Protection Agency, July 2001.

Current EPA regulations under the Clean Air Act and the New Source Performance Standards and Emissions Guidelines specify that many landfills must collect and combust landfill gas (regulated by size of the landfill). To comply with these regulations, landfill owners can either burn the gas off by flaring[23] it or capture the gas by installing a "landfill gas-to-energy" system. (This is discussed in detail later in this report.)

In addition to regulations governing the emission of landfill gases, federal law also regulates the incineration or open burning of waste. Federal law specifically prohibited open burning of MSW at municipal landfills in 1979 (40 CFR 257).[24] The incineration of MSW is strictly regulated by a variety of federal, state, and local policies.

NUMBER OF LANDFILLS. The amount of MSW produced in the United States has risen substantially over the past 50 years, from 88.1 million tons in 1960 to 230 million tons in 1999.[25] On the other hand, the number of landfills has significantly decreased over the last 10 years, from about 8,000 in 1988 to about 2,200 in 1999.[26] Figure 2 shows the decline over the past 14 years; Figure 3 and Table 2 show the number of landfills per state. This decrease in the number of landfills is generally due to stricter regulations imposed by the EPA regarding landfill gas emissions, safety regulations, and content regulations of a landfill. Over the same period, the size of the remaining landfills has grown steadily to accommodate the increased production of MSW.

The number of landfills recorded by the EPA, however, does not take into account all of the individual, and in many cases illegal, dumping sites that were common in the early 1980s. Many businesses, factories, and enterprises had their own dumping sites where they disposed of various types of unregulated wastes. This was a widespread practice before environmental groups began lobbying against such sites and publicizing links between diseases such as cancer and the dumping of hazardous chemicals and toxic wastes that were contaminating water, soil, and air.

THE DEVELOPMENT OF LANDFILL REGULATION.[27] The EPA was established in 1970 after scientists, elected officials, and citizens recognized the need to protect the environment. The new agency was pieced together from programs elsewhere in the federal government, including from the Department of Health, Department of the Interior, and Food and Drug Administration. It was not until 1984 that the EPA gained regulatory authority over landfills. Over the intervening years, various legislative acts have strengthened the EPA's regulatory authority over these sites.

In 1976, Congress passed the Resource Conservation and Recovery Act (RCRA), which gave the EPA the authority to control hazardous waste from the "cradle-to-grave." RCRA covers the generation, transportation, treatment, storage, and disposal of hazardous waste and provides a

---

[23] In this context, *flaring* is the controlled burning of methane collected from a landfill.

[24] "Volume III–Area Sources, Chapter 16, Open Burning," R*evised Final: Emission Inventory Improvement Program Document Series,* Environmental Protection Agency, Section 2.1, January 2001.

[25] *Municipal Solid Waste in 1999: Facts and Figures,* Environmental Protection Agency. Some EPA sources quote this numbers as being closer to 2,300.

[26] *Environmental Fact Sheet, Municipal Solid Waste Generation,* Environmental Protection Agency, 1998.

[27] Information on federal regulations was taken from the EPA website, Major Environmental Laws. http://www.epa.gov/epahome/laws.htm.

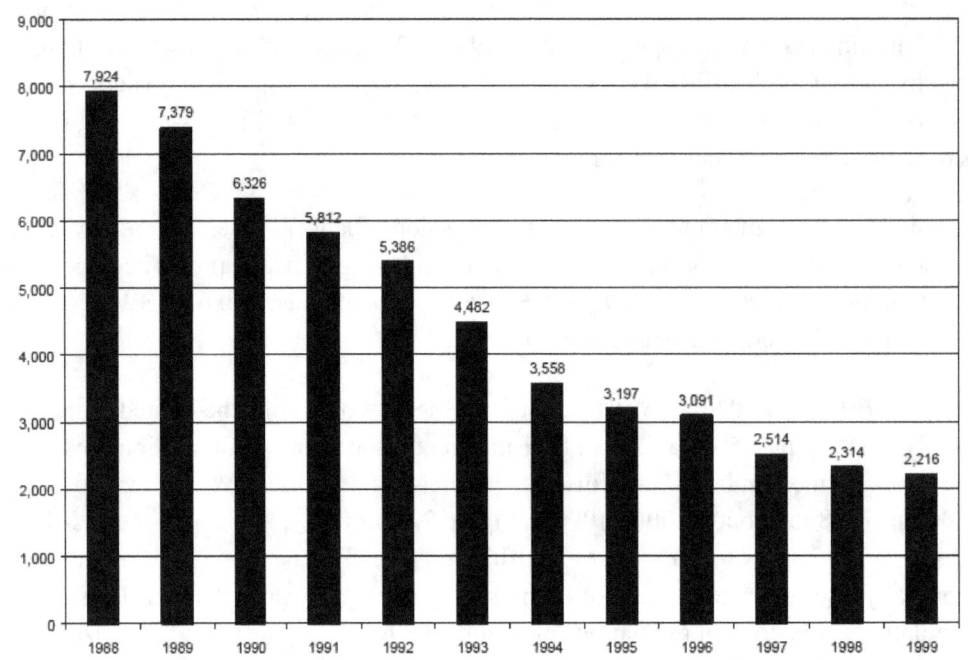

Figure 2. MSW Landfills in the United States, by Year[28]

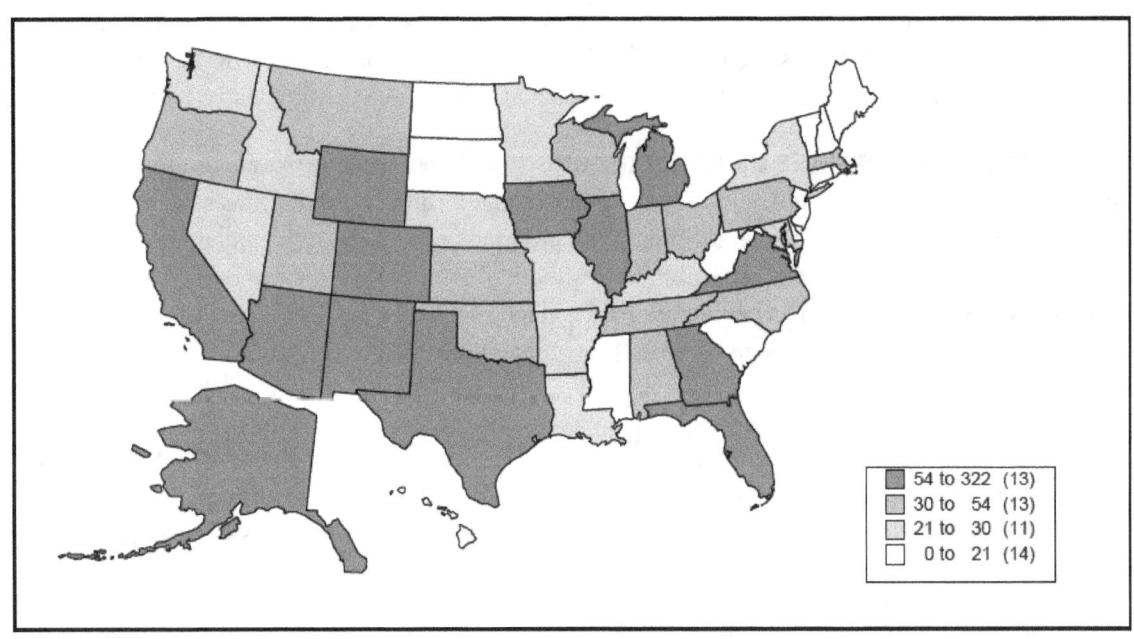

Figure 3. Distribution of Landfills[29]

---

[28] *Municipal Solid Waste in 1999*, op. cit., p. 15.

[29] *BioCycle*, June 1999.

Table 2. Landfills by State[30]

State	Landfills	State	Landfills	State	Landfills
Alabama	30	Kentucky	26	New York	28
Alaska	322	Louisiana	25	Ohio	52
Arizona	54	Maine	8	Oklahoma	41
Arkansas	23	Maryland	22	Oregon	33
California	188	Massachusetts	47	Pennsylvania	51
Colorado	68	Michigan	58	Rhode Island	4
Connecticut	3	Minnesota	26	South Carolina	19
Delaware	3	Mississippi	19	South Dakota	15
District of Columbia	0	Missouri	26	Tennessee	34
Florida	95	Montana	33	Texas	181
Georgia	76	North Carolina	35	Utah	45
Hawaii	8	North Dakota	15	Vermont	5
Idaho	27	Nebraska	23	Virginia	70
Illinois	56	Nevada	25	Washington	21
Indiana	45	New Hampshire	19	West Virginia	19
Iowa	60	New Jersey	11	Wisconsin	46
Kansas	53	New Mexico	55	Wyoming	66

framework for the management of nonhazardous wastes. RCRA focuses only on active and future facilities.

The turning point in landfill regulation and remediation occurred in 1980, first with the Superfund legislation, then by the Hazardous and Solid Waste Amendments (HSWA) in 1984, which finally gave the EPA regulatory authority over landfills.

Technically known as the Comprehensive Environmental Response, Compensation, and Liability Act (CERCLA), the Superfund legislation governs closed and abandoned hazardous material waste sites, provides for the liability of persons responsible for the release of hazardous materials at these sites, and establishes a trust fund to provide for cleanup where no responsible party could be identified.

In 1984, the HSWA amended RCRA. HSWA required the phasing out of land-based disposal of hazardous waste and gave the EPA regulatory authority over landfills. The final major piece of legislation, the Superfund Amendments and Reauthorization Act (SARA), was passed in 1986 as an amendment to CERCLA. SARA increased the participation of states in the Superfund program and expanded the size of the cleanup trust fund.

In recent years, federal, state, local, and private programs have increased the emphasis placed on reducing the production of municipal waste to conserve resources and reduce pollution while delaying the entry of waste into the waste collection and disposal system. "Source

---

[30] Ibid.

reduction" focuses on designing, manufacturing, purchasing, or using materials in ways that reduce the amount or toxicity of trash created.

Some such programs include "pay-as-you-throw," where residents pay for each can or bag of trash they have collected for disposal rather than funding this collection by a flat rate or through the tax base. This provides tangible financial benefits for households that reduce the amount of waste they produce. Other programs target businesses and corporations in an effort to promote waste-reducing manufacturing processes and business practices.[31] The benefits of these practices include a reduction of the combustible material that enters the waste stream. Although MSW facilities will still contain large amounts of combustible materials, this reduction in waste can be a factor in the reduction of landfill fires.

CHARACTERISTICS OF LANDFILL FIRES[32]

Landfill fires fall into one of two categories, surface and underground fires. Depending on the type of landfill and type of fire, landfill fires can pose unique challenges to the landfill/ waste management industry and the fire service. This section addresses the particular challenges and the specific types of fires found in landfill sites and describes their characteristics and causes.

SURFACE FIRES. Surface fires involve recently buried or uncompacted refuse, situated on or close to the landfill surface in the aerobic decomposition layer, generally 1 to 4 feet in depth.[33] These fires can be intensified by landfill gas (methane), which may cause the fire to spread throughout the landfill.

Surface fires generally burn at relatively low temperatures and are characterized by the emission of dense white smoke and the products of incomplete combustion. The smoke includes irritating agents, such as organic acids and other compounds. When surface fires burn materials such as tires or plastics, the temperature in the burning zone can be quite high. Higher temperature fires can cause the breakdown of volatile compounds, which emit dense black smoke. Surface fires are classified as either accidental or deliberate.

Surface fires include the following:

- *Dumping of undetected smoldering materials into the landfill.* Hot load fires are caused by the disposal of refuse that is still burning on arrival to the landfill (e.g., cleared brush).

- *Fires associated with landfill gas control or venting systems.* Landfill gas control systems can themselves pose a fire hazard. Landfill gas (predominantly methane) can be

---

[31] "Source Reduction and Reuse," Environmental Protection Agency, April 23, 2002. http://www.epa.gov/epaoswer/non-hw/muncpl/sourcred.htm.

[32] Much of this section represents a synopsis of a report prepared for the New Zealand Ministry of the Environment. The report, *Landfill Guidelines: Hazards of Burning at Landfills*, was published in December 1997.

[33] E-mail correspondence with Todd Thalhamer, California Integrated Waste Management Board.

ignited as it escapes from the vents or from leaks in the collection pipe network. Excessive gas extraction can also be a fire cause. The vacuum created by excessive extraction can increase the airflow and thereby increase the oxygen level in the land-fill, which can cause underground fires (as discussed further in the following section).

- *Fires caused by human error on the part of the landfill operators or users.* Landfill operators and users can cause fires through careless smoking on the landfill, which can ignite waste or landfill gas. Also, as some hazardous substances can ignite when mixed, operators must take care to prevent the dumping of reactive materials into the landfill.

- *Fires caused by construction or maintenance work.* Fires can occur while construc-tion and maintenance takes place, including fires caused by sparks from vehicles used in the landfill (dump trucks, bulldozers, backhoes, etc.). A surface fire could also be ignited when drilling or while driving metal pipes through layers of buried waste if a hard object buried in the landfill is struck. Usage of welding or electrical equipment on site poses a fire hazard, due especially to the increased presence of methane gas.

- *Spontaneous combustion of materials in the landfill.* The mixing of certain materials in a landfill can result in spontaneous combustion. Even in small quantities, some chemicals can ignite if exposed to one another. Also, some materials, such as oily rags, can spontaneously combust under certain conditions. Spontaneous combustion can also result from bacterial decomposition, which is discussed in more detail later in this section.

- *Deliberate fires, which are used by the landfill operator to reduce the volume of waste.* Landfills contain refuse such as dry garden waste, grass, leaves, and branches. Sometimes these materials are deliberately set on fire to reduce refuse volumes, reduce operating costs, and increase a landfill's operating life. This is an accepted practice under strictly controlled conditions.[34] Uncontrolled, these deliberate fires could escalate into larger fires, cause explosions, or create hazardous products from the ash and residue burned.

- *Deliberate arson fires, which are set with malicious intent.* Arson is a serious prob-lem in the United States; therefore, it is not surprising that landfills are targets for malicious fires.

UNDERGROUND FIRES. Underground fires in landfills occur deep below the landfill surface and involve materials that are months or years old.[35] These fires are generally more diffi-cult to extinguish than surface fires. Underground fires also have the potential to create large

---

[34] This controlled combustion at landfills is regulated by U.S. Code of Federal Regulations, 40 CFR 60 (Title 40 – Protection of Environment Chapter I – Environmental Protection Agency. Part 60 – Standards Of Performance For New Stationary Sources).

[35] This report addresses operating landfills. Closed landfills are subject to a variety of restrictions on future develop-ment, maintenance, etc. It would be difficult to determine the frequency of fires in closed landfills because such sites are likely to be coded in NFIRS according to their property use at the time of the fire (e.g., open land, park, golf course).

voids in the landfill, which can cause cave-ins of the landfill surface. Further, they produce flammable and toxic gases (such as carbon monoxide) and can damage leachate containment liners and landfill gas collection systems.

The most common cause of underground landfill fires is an increase in the oxygen content of the landfill, which increases bacterial activity and raises temperatures (aerobic decomposition). These so-called "hot spots" can come into contact with pockets of methane gas and result in a fire. Of particular concern with these long-smoldering, underground fires is the fact they tend to smolder for weeks to months at a time. This can cause a build up of the byproducts of combustion in confined areas such as landfill site buildings or surrounding homes, which adds an additional health hazard.

Underground fires are often only detected by smoke emanating from some part of the landfill site or by the presence of carbon monoxide (CO) in landfill gas. In the event of an underground fire, CO may be present at toxic levels near the landfill's surface. Generally an underground fire can be confirmed by:[36]

- Substantial settlement over a short period of time.

- Smoke or smoldering odor emanating from the gas extraction system or landfill.

- Elevated levels of CO in excess of 1,000 parts per million (ppm).

- Combustion residue in extraction wells or headers.

- Increase in gas temperature in the extraction system (above 140°F).

- Temperatures in excess of 170°F.

To confirm a subsurface fire using CO, the results must be acquired through quantitative laboratory analysis (using portable monitors may result in artificially high concentrations). In California, levels of CO in excess of 1,000 ppm are considered a positive indication of an active underground landfill fire. Levels of CO between 100 and 1,000 ppm are viewed as suspicious and require further air and temperature monitoring. Levels between 10 and 100 ppm may be an indication of a fire but active combustion is not present.[37]

HEALTH EFFECTS OF LANDFILL FIRES. In addition to the burn and explosion hazards posed by landfill fires, smoke and other byproducts of landfill fires also present a health risk to firefighters and others exposed to them. Smoke from landfill fires generally contains particulate matter (the products of incomplete combustion of the fuel source), which can aggravate pre-existing pulmonary conditions or cause respiratory distress. As with all fires, those in landfills produce toxic smoke and gases. The danger and level of toxicity of these gases depend on the length of exposure one has to them and on the type of material that is burning.

---

[36] *Response to Landfill Fires Guidance Document*, California Integrated Waste Management Board, Internal Bulletin 2001.
[37] Ibid.

Underground fires can result in CO levels in excess of 50,000 ppm—the Occupational Safety and Health Administration (OSHA) permissible exposure limit for CO is 50 ppm. OSHA standards prohibit worker exposure to more than 50 parts of the gas per million parts of air averaged during an 8-hour time period. Carbon monoxide is harmful when breathed because it displaces oxygen in the blood and deprives the heart, brain, and other vital organs of oxygen, which can cause permanent damage or death.[38]

Another serious concern in landfill fires is the emission of dioxins. Accidental fires at landfills and the uncontrolled burning of residential waste are considered the largest sources of dioxin emissions in the United States.[39] The term *dioxins* refers to a group of chemical compounds with similar chemical and biological characteristics that are released into the air during the combustion process. Dioxins are also naturally occurring and are present throughout the environment. However, exposure to high levels of dioxins has been linked to cancer, liver damage, skin rashes, and reproductive and developmental disorders.[40]

EXTINGUISHING LANDFILL FIRES

This section is not intended to address or recommend specific tactical approaches for landfill firefighting. It is important to note that the different dynamics, characteristics, and regulations of landfills and the fires that occur in them suggest that tactics need to be determined on a case-by-case basis depending on the materials buried, which materials have ignited, depth of the fire, and the fire's ignition source. This section explores some of the challenges posed in the suppression of landfill fires.

WIND/WEATHER. Wind and inclement weather can increase the health hazards for firefighters operating on the fireground (e.g., in extremely hot or cold weather) and can directly affect fire spread.

WATER SUPPLY. The use of water to suppress landfill fires is controversial. The application of large volumes of water may actually exacerbate a fire by contributing to the process of aerobic decomposition. Further, adding water to the landfill creates additional leachate, which may overwhelm the leachate collection system in the landfill (if one exists). If the collection system is overwhelmed, the additional leachate could contaminate ground and surface waters surrounding the landfill. Depending on the landfill's location, there might not be an adequate supply of water available for fire suppression. Firefighters may have to establish a water supply using tankers and nearby static water sources (e.g., lakes, reservoirs).

---

[38] *OSHA Fact Sheet, Carbon Monoxide Poisoning*, U.S. Department of Labor, Occupational Safety and Health Administration, 2002. http://www.osha.gov/OshDoc/data_General_Facts/carbonmonoxide-factsheet.pdf

[39] *Questions and Answers About Dioxins*, Environmental Protection Agency, July 2000, p. 6. http://www.epa.gov/ncea/pdfs/dioxin/dioxin%20questions%20and%20answers.pdf.

[40] Idem, p. 4.

Foam is an important consideration in landfill fire suppression. There are two primary types of firefighting foam. Class A foam is a special formulation of hydrocarbon surfactants. These surfactants reduce the surface tension of water, which provides for better water penetration and increased effectiveness. When aerated, Class A foam coats and insulates fuels, protecting them from ignition. Class B foam is used to extinguish fires involving flammable and combustible liquids. It is also used to suppress vapors from unignited spills of these liquids.[41] As with all fires, there are advantages and disadvantages to using foam during fire suppression operations on landfills. The on-scene incident commander makes the decision to use foam based on the specific tactical situation at hand.

MULTI-AGENCY RESPONSE. A major landfill fire will likely require the expertise of personnel from multiple agencies (e.g., the EPA, Department of Natural Resources). Some fire departments have Standard Operating Procedures in place that define all landfill fires as hazardous materials incidents, which require a specialized response. To ensure that all personnel (regardless of their agency affiliation) are operating according to the same plan, landfill fires require a strong Incident Command System.

PERSONNEL SAFETY. Fires, particularly those underground, can undermine the integrity of the landfill, which could cause a collapse under the weight of landfill employees, firefighters, or equipment. Such a collapse could necessitate a confined space, trench, or other type of technical rescue operation in addition to fire suppression.

Given the potential adverse effects of exposure to burning landfill contents or the smoke produced by a landfill fire, personnel may have to use specialized personal protective equipment, which may be difficult to obtain.

ACCESS TO AND MANEUVERABILITY OF HEAVY EQUIPMENT. To access waste below the landfill surface or move burning waste away from the landfill, it may be necessary to use heavy equipment such as bulldozers. Landfill operators may already own this equipment and have staff trained in its use. If not, this equipment will need to be located and brought to the fireground. If a fire affects the structural stability of a landfill, operating heavy equipment on the landfill surface would be dangerous. Finally, depending on the landfill's location and design, operating heavy equipment on the site could be quite difficult.

LOGISTICS. As with any protracted fire suppression operation, Incident Commanders at landfill fires must address a variety of logistical concerns to facilitate operations. These include rotating personnel on a regular basis, compensating personnel for overtime spent operating at the landfill or filling in at fire stations in the jurisdiction, keeping firefighters on the landfill hydrated and fed, and, keeping records for future reimbursement. (Depending on the nature and location of the incident, local fire departments can seek reimbursement from the federal government or the landfill operator for costs associated with fire suppression.)

ENVIRONMENTAL IMPACT. The smoke and runoff from landfill fires can be dangerous to those living in the area and to the environment. It is important that air and water quality issues

---

[41] *Essentials of Firefighting 4th Edition*, International Fire Service Training Association, 2001, p. 500.

be addressed early in a fire suppression operation to prevent contamination as much as possible. As mentioned earlier, water used to suppress a landfill fire can overwhelm a facility's leachate collection system, if one exists (older facilities may have been constructed prior to regulations requiring leachate collection systems).

LANDFILL CONTENTS. Fires occurring in landfills where hazardous wastes are buried can be particularly difficult. In past years, illegal dumping of hazardous and toxic materials in landfills and other dumping sites was relatively common. When a fire occurs and rescue workers have wrong or misleading information about the buried contents (e.g., illegal or unknown toxic or radioactive wastes), the fire suppression operation can be extremely dangerous.

Although not a landfill fire, the Wade Dump fire in February 1978 clearly illustrates the dangers posed by fires involving unknown hazardous materials. Firefighters responded to a suspected tire fire at an abandoned rubber shredding plant on the Delaware River outside of Philadelphia. They were unaware that the property's owner and namesake, Melvin Wade, had transformed the plant into one of the most toxic hazardous waste dumpsites in U.S. history. By the night of the fire, more than 3 million gallons of cyanide, benzene, toluene, and other chemicals were stored on the site—plus thousands of junk tires. The burning chemicals produced multicolored smoke and noxious fumes, which alerted firefighters to the unusual nature of the fire they were fighting. Intensified by chemicals and other fuels, the fire raged for hours. Drums of chemicals exploded, injuring firefighters and even damaging fire trucks. As the night progressed, firefighters and other emergency workers noticed that the chemicals were dissolving their protective gear and making it difficult for them to breathe; more than 40 firefighters were sent to a nearby hospital for treatment. Over the past 20 or more years, dozens of those who were present at the Wade Dump fire have become ill, and many have died from cancers and other diseases. Melvin Wade and others responsible for creating the toxic site were found criminally responsible for their actions.[42]

LANDFILL FIRES: STATISTICAL ANALYSIS

Data from the National Fire Incident Reporting System (NFIRS) does not include MSW landfills as a fixed property use category. Rather, the NFIRS data set includes a category for "dump or sanitary landfill: included are refuse disposal areas, trash receptacles, and dumps in open ground" (NFIRS Fixed Property Use code 932). Although this definition is broader than the definition of a landfill, it is the closest match available in NFIRS. As such, despite the broader definition, this section refers to these fires as *landfill fires* for the sake of clarity.

Based on extrapolation of the NFIRS data, each year in the United States an average of 8,400 landfill fires are reported to the fire service. This represents less than a half percent of all reported fires. Undoubtedly, some landfill fires go unreported because they burned undetected or they were on private property and extinguished by the landfill operator. Reported fires are responsible for less than 10 civilian injuries, 30 firefighter injuries, and between $3 and

---

[42] This paragraph is a synopsis of an investigative report published by the *Philadelphia Inquirer* in April 2000.

$8 million in property loss each year.[43] Deaths (civilian or fire service) are rare in these fires; since NFIRS represents a sample of data, it may be that fatalities occurred during the study period and were not reported or captured in the data.

TYPE OF LANDFILL FIRES. Table 3 shows the five types of fires that occur on landfills. The prevalence of refuse fires is not surprising, but it is interesting that other types of fires occur on landfill properties. Vehicle fires involve dump trucks, compactors, and other vehicles commonly found in landfills. Brush fires may occur when landfill fires spread to the surrounding lands. Structure fires at landfill sites probably involve small offices or other facilities constructed for the landfill staff.

Table 3. Types of Fires Occurring on Landfills[44]

Type of Fire	Percent of Fires
Refuse	77
Trees, brush, grass	12
Outside structure, where material burning has value	6
Vehicle	4
Structure	1

CAUSES OF LANDFILL FIRES. Over half of the landfill fires reported to NFIRS have no information available as to the primary ignition factor. This makes it particularly difficult to accurately pinpoint the cause of landfill fires. Of those fires with reported ignition factors, nearly 40 percent are of an incendiary or suspicious nature. Another 20 percent are attributed to lit or smoldering materials that have been abandoned or discarded, which include cigarettes, matches, or ashes that were discarded without being properly extinguished. Spontaneous heating accounts for about 5 percent of landfill fires. Other leading factors influencing fire ignition include rekindling from a previous fire and inadequate control of open fires.

WHEN LANDFILL FIRES OCCUR. Landfill fires occur most often between March and August. This half-year period accounts for nearly 60 percent of landfill fires, with the peak (11 percent) occurring in July (Figure 4). This monthly incidence of fires generally applies to the major causes of landfill fires (incendiary/suspicious and smoldering materials). Rekindled fires and spontaneous ignition fires, however, are exceptions. Rekindled fires have a peak period in April and May that accounts for one-third of these fires with an additional peak in July (15 percent). Landfill fires that result from spontaneous combustion gradually increase as the weather warms, dropping in September. The peak period, however, occurs in October and November, when 22 percent of the spontaneous combustion fires occur. Figure 5 illustrates the incidence of spontaneous combustion fires by month.

---

[43] National estimates are based on NFIRS data (1996–1998) and the National Fire Protection Association's (NFPA) annual survey, *Fire Loss in the United States*.

[44] U.S. Fire Administration NFIRS data (1996–1998).

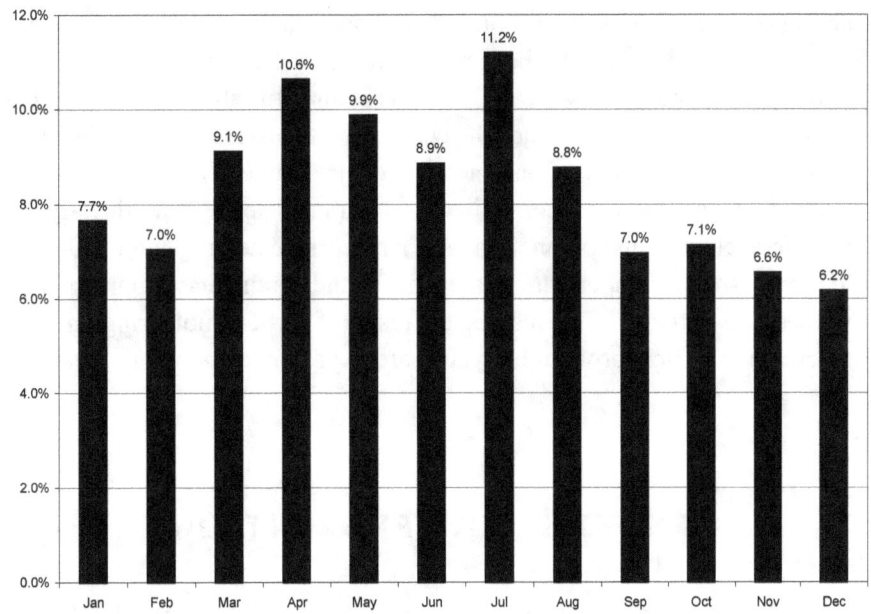

Figure 4. Incidence of Landfill Fires by Month[45]

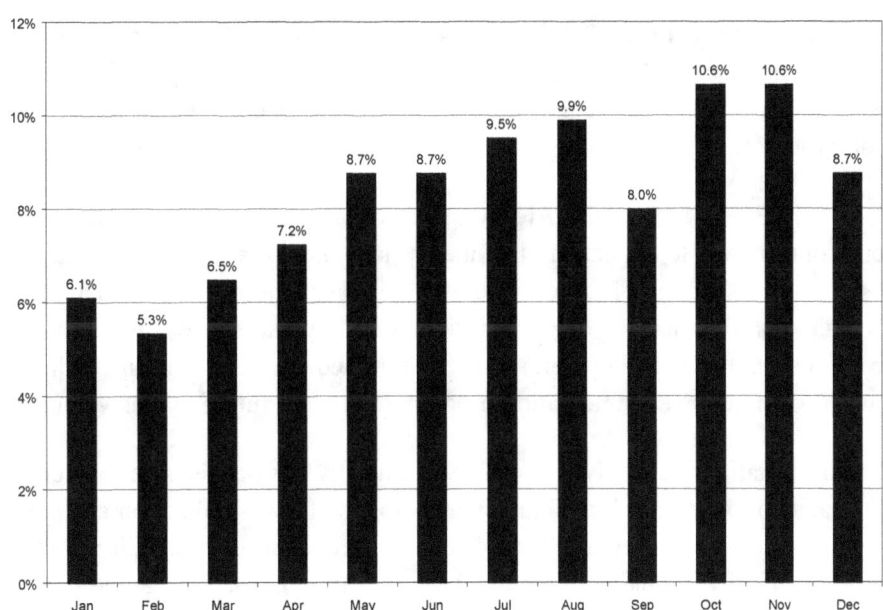

**Figure 5. Incidence of Spontaneous Combustion
Landfill Fires by Month**[46]

---

[45] Ibid.
[46] Ibid.

The spring peaks in rekindled fires and the fall peaks in spontaneous combustion fires may result from increased winds during these months as many landfills may have inadequate caps (particularly if they use alternate daily covers) to prevent air infiltration. Inadequate caps can allow large volumes of air to enter the landfill, accelerating the oxidation reaction. The air intrusion is due, in part, to the differential in barometric pressure between the landfill and the atmosphere. This condition occurs most frequently in the late fall and spring with the large, naturally occurring atmospheric changes in conjunction with land surface heating and cooling. The increased oxidation raises the temperature in the landfill and can increase spontaneous combustion events. Some of the rekindled fires may be the result of earlier smoldering underground fires that, with the increase in airflow brought by winds, are oxygenated enough to break through to the surface.[47]

LANDFILL FIRE PREVENTION

Fire prevention can reduce property damage, injury, health, and environmental hazards of landfill fires. The cost of prevention is usually much less expensive than the cost of fighting and cleaning up a fire. In many cases, particularly for larger landfills, fire prevention activities are required by law. This section outlines some of the principal methods in landfill fire prevention.

LANDFILL MANAGEMENT. Effective landfill management is a vital key to efficient landfill fire prevention tactics. Management measures include prohibiting all forms of deliberate burning, thoroughly inspecting and controlling incoming refuse, compacting refuse buried to prevent hot spots from forming, prohibiting smoking onsite, and maintaining good site security.

METHANE GAS DETECTION AND COLLECTION. Landfill gas emissions can be a hazard to the environment and to the health of residents surrounding landfill sites. Methane gas, a flammable gas, can present a fire hazard. Federal regulations require all MSW landfill operators to monitor the emission of methane on a quarterly basis. If methane levels in or around the landfill become explosive, the landfill operator must take immediate steps to mitigate the danger. The operator must also implement a remediation program to prevent future explosive buildups.[48]

Federal regulations currently require MSW landfills that opened after November 8, 1987, and have a capacity of over 2.5 million cubic meters to install a gas collection and control system.[49] These regulations, however, affect only about 4 percent of operating landfills in the United States as the vast majority of landfills do not have such a large capacity.[50] Some states, however, (e.g., California) have stricter regulations for gas collection systems, which affect a higher percentage of facilities; these jurisdictions may include closed facilities as well.

---

[47] E-mail correspondence with Dr. Tony Sperling and Todd Thalhamer.

[48] U.S. Code of Federal Regulations, 40 CFR 258.23 (Title 40–Protection of Environment Chapter I–Environmental Protection Agency. Part 258 – Criteria for Municipal Solid Waste Landfills).

[49] U.S. Code of Federal Regulations, 40 CFR 60.33c (Title 40–Protection of Environment Chapter I–Environmental Protection Agency. Part 60 – Standards of Performance for New Stationary Sources).

[50] *Air Rule for Municipal Solid Waste Landfills*, Environmental Protection Agency, January 10, 2002. http://www.epa.gov/reg3artd/airregulations/ap22/landfil2.htm.

Methane gas collection systems actively remove landfill gas using gas recovery wells and vacuum pumps with an interconnected pipe network. Operators must take care to ensure the system is not overdrawn, which can lead to fire ignition. Once the gas is collected, landfill owners/ operators have two choices: (1) burn off the gas (flaring); or (2) convert the gas to an energy commodity.

Flaring. Burning landfill gas is the method most large landfills use (as opposed to the more costly waste-to-energy projects). Burning the landfill gas converts methane to carbon dioxide, which not only is less harmful to the environment, but also destroys the components of landfill gas that cause odor, stress vegetation, create smog, and increase the risk for fire or explosion.

Shallow gas venting trenches or gas venting pipes can also be installed in the landfill's surface. These vents allow gas from interior regions of the landfill to escape naturally to the surface where flares can burn off the gas.

Converting Landfill Gas to Energy. The conversion of landfill gas to energy turns this landfill byproduct into a marketable resource. The converted gas can be used to generate electricity, heat, or steam. According to the EPA, landfill gas is the only renewable energy source that, when used, removes pollution from the atmosphere.[51] By converting the landfill gas to energy, the harmful emissions causing global warming are removed from the air and converted to a useful form such as electricity to power a home. Reducing landfill gas emissions is imperative as it reduces local ozone levels and smog formation while simultaneously decreasing explosion and fire risks and unpleasant odors produced by the landfill.[52]

As of September 2001, the EPA estimates that there were more than 335 landfill gas recovery and utilization projects operating in the United States; another 500 landfills are considered good candidates for future program development.[53]

CASE STUDIES

A sample of landfill fires throughout the world sheds light on the landfill fire problem. Waste disposal practices and the regulation of landfill sites are similar in the comparison countries. Landfill fires have been investigated and studied in more detail in several countries outside the continental United States. The concluding portion of this section contains brief synopses of interviews and media reports detailing landfill fires in the United States and the lessons that were learned from them.

---

[51] Landfill Methane Outreach Program, *Frequently Asked Questions*, Environmental Protection Agency, updated June 5, 2001. http://www.epa.gov/lmop/faq.htm.

[52] Ibid.

[53] Landfill Methane Outreach Program, *Current Projects and Candidate Landfills*, Environmental Protection Agency, January 10, 2002. http://www.epa.gov/lmop/projects.htm.

FINLAND.[54] An experimental study that sheds significant light on methods of extinguishing landfill fires was conducted in Finland in 1993. The study was conducted in two parts: a questionnaire was distributed to landfill operators throughout Finland, and an experimental landfill was constructed with similar characteristics to an MSW landfill. To determine the most effective methods for extinguishing landfill fires, an underground fire was ignited and allowed to burn in the experimental landfill. The fire was extinguished by smothering it with soil and dousing it with water.

From the questionnaires, the study determined that most landfill fires are small and tend to be of short duration. It concluded that using soil and water to extinguish the fires was insufficient and that a potentially significant factor in landfill fires is the improper compaction of waste in the landfill. The study suggested that one way to prevent landfill fires is to sufficiently compact all waste buried in the landfill site. Only one-quarter of the fires reported to the study team were underground; those fires were particularly difficult to extinguish and tended to last over 2 months. In fact, for underground fires, it was found that covering the smoldering refuse with layers of soil actually prolonged some fires. Another serious concern raised in the study was that by using water to extinguish landfill fires, the runoff could contaminate the surrounding soil and ground water.

Ultimately, based on both the questionnaire and the experimental landfill, the study concluded that the most effective way to suppress landfill fires is by digging out the burning material and cooling it with water, soil, or snow.[55]

CANADA.[56] In November 1999, a fire ignited at the Delta Shake and Shingle Landfill, a C&D landfill outside Vancouver, British Columbia. Although smoke and steam had been emanating from the landfill for weeks, the fire was finally discovered when flames broke through the landfill surface. The landfill operator originally attempted to extinguish the fire without fire department assistance; his efforts only served to exacerbate the fire. After several weeks, residents began to complain about the smoky haze hovering over Vancouver, and officials were concerned about air and water contamination from the suppression efforts. Ultimately, local officials declared a state of emergency and requested assistance from both the private sector and the provincial government.

To contain the fire and starve it of oxygen, officials covered the burning materials with a thick layer of refuse. Next, they determined that although using high-pressure water worked to extinguish the surface fire, it did not extinguish the burning refuse underground. To increase the water's effectiveness, firefighters misted the water and added Class A foam. Once the fire was contained, the firefighters used heavy machinery to excavate burning materials and move them to

---

[54] Ettala et al., "Landfill Fires in Finland," *Waste Management & Research* (1996) 14, pp. 377-384.

[55] Other landfill fire suppression professionals, however, have found that landfill fires can be extinguished by excavating and extinguishing the burning debris layer-by-layer using soil and a suppressant agent, or simply by temporarily shutting down the gas extraction system.

[56] Sources for this section: "Landfill Fire in Delta Gets Provincial Emergency Funding," British Columbia Ministry of Environment, Lands, and Parks. Press Release 330-30:ELP99/00-340, November 30, 1999. Sperling, Tony. *Extinguishing the Delta Shake and Shingle Landfill Fire: Case Study*, Sperling Hansen Associates, January 18, 2002. http://www.landfillfire.com/delta1.html.

areas offsite where they could be fully extinguished. Firefighters used infrared technology to determine which loads were "hot" and required extinguishment and which ones were cool enough to be left alone. After the materials were fully extinguished using foam and water, they were returned to the reconstructed landfill.

A private contractor involved in the suppression effort summarized the following as lessons learned from this fire:

- Soil berms are effective at containing fire spread.

- Trenches that do not fully penetrate the refuse pile are ineffective; trenches should only be excavated if they penetrate the full thickness of the refuse to inert material.

HAWAII. In the late 1990s, fires in legal and illegal landfills were a serious concern for officials on all of the Hawaiian Islands. In July 1996, a fire at an illegal dumpsite in Lualualei, Oahu, attracted government and media attention. The site contained municipal waste, C&D debris, and hazardous materials. After explosions involving gas cylinders or drums, the State Department of Health hired a hazardous waste contractor to remove drums containing chemicals and some hazardous waste. Despite the attention, government officials had difficulty shutting down the dumpsite, as the property changed hands over the years and the cost of cleaning up the site exceeded the land's value.[57]

In January 1998, an odd odor at a C&D landfill in Ma'alaea led to the discovery of an underground fire.[58] Efforts to extinguish the fire with carbon dioxide were unsuccessful and, while the fire was contained, it smoldered for months.

Hawaii has less rigorous air quality standards than other areas of the United States because of its tradewinds, low population density, and isolation. Contractors are allowed to burn brush before depositing it in landfills. This practice decreases the waste volume and amount they are charged for using the landfills. Burned material goes through two inspection sites to check for "hot loads." In the Ma'alaea fire, it appears the ignition source was a smoldering palm tree. Palm trees are spongy inside and, though the outside may have appeared cool, the inside was still simmering. Once inside the landfill, the tree continued to smolder until it ignited surrounding waste.

Although relatively small, the fire sparked a debate involving the landfill operator, EPA, and different divisions of the Department of Health. The debate revealed that there were no regulations on methods to control landfill fires. This motivated government officials to develop guidelines that address underground fires and study the health effects of landfill fires. Also, the fire emphasized the need to thoroughly inspect suspected hot loads to ensure that smoldering materials do not accidentally enter the landfill.

---

[57] "State Health Department To Close Illegal Dump in Lualualei," *Environment Hawaii*, Volume 11, Number 3, September 2000.

[58] "Ma'alaea Landfill Sparks State Effort To Develop Guidelines," *Environment Hawaii*, Volume 9, Number 4, October 1998.

OTHER EXAMPLES. The following examples were taken from media reports and interviews with fire officials in the affected jurisdictions. These examples shed light on firefighting tactics and local concerns associated with landfill fires.

Fairfax County, Virginia.[59] Fairfax County Fire Station 19 (Lorton) has two landfills within its call range. In November 2000, a fire broke out at the I-95 Landfill, near Lorton, VA. A 250-foot by 50-foot pile of debris, consisting of trees, stumps, and mulch, was ignited. Firefighters used water and foam to control and extinguish the fire. A fire technician who participated in the suppression effort stated that the most important tactic used in the fire was having firefighters and machinery overhaul the burning or smoldering areas to ensure that the fire did not rekindle.

Cumberland County, North Carolina.[60] In July 1998, flames at a landfill sent plumes of smoke over a large area. Firefighters were forced to contain the fire and let it burn since it was too hot for water to extinguish it effectively. An estimated 26 trailer loads of mulch were in the landfill. The mulch was very finely packed, the heat remained at the core, and water would not have cooled or extinguished the fire. Firefighters assured the fire did not spread to nearby tire piles by digging a ditch all around the fire, containing it. The fire burned itself out after several weeks.

Montezuma County Landfill, Colorado.[61] In June 2001, smoke from this 6-acre fire spread high over the Montezuma Valley. The 320-acre landfill was filled with compressed, baled trash and municipal and industrial waste.[62] Attempts were made to douse the fire with water, but they were ineffective. State landfill officials and other experts decided the best way to attack the blaze was to remove the smoldering bales of refuse, break them apart, and extinguish them individually. The cause of the fire was not determined. Landfill officials reported that confining the fire and smothering it proved to be the most effective method of extinguishing it.

Danbury, Connecticut[63] In 1996 and 1997, numerous underground landfill fires occurred at the Danbury city landfill. These fires were caused by spontaneous combustion of decomposing waste and were rekindled and continued smoldering underground over 18 months. Different underground "hotspots" increased the intensity of landfill odors. These fires in the 47-acre landfill were the subject of extensive media coverage and residential complaints. As elsewhere, water was ineffective in extinguishing these fires, and its use added to the stench, causing additional citizen complaints. Residents filed lawsuits for damages caused by exposure to hydrogen sulfide gas from the smoke. As a result of the lawsuits, the landfill was forced to close. A 40-foot high permanent flare had to be installed to burn off landfill gas and reduce the odors.

Bend, Oregon.[64] A youth fell into a burning sinkhole on the site of an old landfill and suffered third-degree burns across 30 percent of his body. The youth and his friend had noticed a thin trail of smoke coming from the ground while walking home and went to investigate. There

---

[59] Telephone interview with David Sweedland, Technician, Fairfax County Station 19, and *I-95 Landfill Debris Fire*, Fairfax County Fire and Rescue Department News Release, November 7, 2000.

[60] *Landfill Fire Continues To Burn*, WRAL 5 Cumberland County News, July 30, 1998.

[61] "Landfill Fire Fills Valley With Smoke," *Cortez Journal*, June 19, 2001.

[62] Telephone interview with Montezuma County Landfill official.

[63] *The News-Times*, Danbury, CT, December 1996–October 1997.

[64] "Youth Slips Into Burning Bend Sinkhole," *The Oregonian*, December 28, 1991.

was a small hole at the surface. While investigating the hole, the ground collapsed around the youth. The sinkhole was on a parcel of park district land on the outskirts of Bend, Oregon. The former landfill was owned by the county, and the land was later given to the park district. The original dump was used for wood waste. The decomposing waste smoldered and ignited through spontaneous combustion. Burned out pockets caused the landfill's earthen cover to weaken and collapse. Most of the problem areas were along the edges of the landfill where the earthen cap was the thinnest. The park district originally planned to put children's baseball fields on an unused portion of the old landfill, but reconsidered after conferring with the local Department of Environmental Quality.

Colerain Township, Ohio.[65] In 1996, the Colerain Township landfill experienced a major landslide that filled a nearby limestone quarry with acres of landfilled waste. The quarry was being excavated to hold additional waste in the landfill site when the landslide occurred. The area that had collapsed was dangerous; garbage was exposed and equipment was buried underneath, which made removal of the waste dangerous. The landfill officials could not move equipment to the site due to enormous voids in the exposed area; they feared bulldozers would be swallowed into the pile.

A series of four fires subsequently ignited, covering a 35-acre area. The first was a small 100-square-yard fire ignited by lightning. The second fire was as a result of combustion of decomposing waste and lasted 7 days covering a 20-acre area. Firefighters used pumped water and heavy equipment to tear down the fire area and then smothered it with dirt. Fifteen to 20 million gallons of water were used in the 7-day period. The last two fires were also a result of spontaneous combustion, but they were smaller in size. Water and heavy equipment were used to extinguish these two fires as well. Ultimately, restoring the landfill took approximately 2 years to complete.

San Bernardino County, California[66] In 1999, funding was approved for the cleanup of a smoldering fire at an illegal dumpsite in Cajon Pass. The illegal dumpsite had been in operation for about 3 years. At the time of the fire, the dumpsite contained 200,000 cubic yards of waste, which filled an area about 60 feet high and 450 feet long. Most of the waste consisted of rubble, telephone poles, railroad ties, whole trees, shrubs, and large stumps. About 80,000 cubic yards (60,000 tons) were organic wastes, which spontaneously ignited, causing the fire. The smoldering fire posed a significant risk to nearby residences, wildlands, power lines, and railroad tracks, and it threatened serious water contamination. Agencies from the state and local level were involved in the funding effort.

---

[65] Telephone interview with Ohio Colerian Township Dept. of Fire and EMS Fire Chief Bruce Smith.

[66] *State Waste Board Approves Funding for Cajon Pass Dump Cleanup,* California Integrated Waste Management Board, May 27, 1999, 99-053. http://www.ciwmb.ca.gov/pressroom/1999/may/nr053 htm.

CONCLUSION

Landfill fires are not common occurrences. When they do occur, however, they tend to attract a great deal of public attention and challenge the fire service. Illegal dumping continues to be a problem for regulatory agencies and the fire service. Illegal sites are particularly hazardous to firefighters, because the firefighters may be unaware of the presence or nature of chemicals or other toxic substances involved in the fire. Landfill fires in regulated facilities also challenge Incident Commanders, who must make a series of tactical decisions in a situation far different from that found at a "normal" structure fire.

Closed landfills are another area of concern, from both a regulatory and a fire service perspective. By federal law, landfill operators must commit to maintaining a landfill for at least 30 years after it has closed. Landfills continue to emit methane and other dangerous gases even after they are closed. As a result, buildings constructed on former landfills are often required to have automatic methane detectors, which sound an audible alarm in the event that methane levels become unsafe. Construction on closed landfills must not damage the final cover or the existing liners and leachate collection system. The true implications of closed landfills are not clear, largely because, for data collection purposes, these sites are likely coded not as landfills but as the property use at the time of an incident (fire, explosion, etc.).

Through EPA regulation and cleanup efforts of landfills, landfill fires are less likely to contain toxic chemicals than they were decades ago. Also, fire departments are gaining the experience to more efficiently and safely extinguish the fires that occur. Working in conjunction with the public and landfill operators, the fire service can reduce the occurrence of landfill fires, thereby better protecting the public, the environment, and emergency responders.

www.ingramcontent.com/pod-product-compliance
Lightning Source LLC
Chambersburg PA
CBHW081415170526
45166CB00010B/3356